60-200

60-200

# WILLIAM DAVID COOLIDGE

*By the same author*

---

H. A. Liebhafsky, H. G. Pfeiffer, E. H. Winslow, and P. D. Zemany, *X-Rays, Electrons, and Analytical Chemistry: Spectrochemical Analysis with X-Rays*, 2nd edition, 1972.

H. A. Liebhafsky and E. J. Cairns, *Fuel Cells and Fuel Batteries: A Guide to Their Research and Development*, 1968.

H. A. Liebhafsky, H. G. Pfeiffer, E. H. Winslow, and P. D. Zemany, *X-Ray Absorption and Emission in Analytical Chemistry*, 1960.

William David Coolidge in his ninety-sixth year

# William David Coolidge

---

## A Centenarian and His Work

HERMAN A. LIEBHAFSKY

*General Electric Company* 1934–1967
*Texas A&M University* 1967–

A Wiley-Interscience Publication

JOHN WILEY & SONS, New York ● London ● Sydney ● Toronto

Published by John Wiley & Sons, Inc.
Copyright © 1974, by Herman A. Liebhafsky

**Library of Congress Cataloging in Publication Data:**
Liebhafsky, H A
William David Coolidge: a centenarian and his work.

"A Wiley-Interscience publication."
Includes bibliographical references.
1. Coolidge, William David, 1873-
2. Tungsten.

TK140.C65L53    541'.3'0924 [b]    74-11233
ISBN 0-471-53430-7

Printed in the United States of America

10 9 8 7 6 5 4 3 2 1

To the memory of

---

LAURENCE A. HAWKINS

*Executive Engineer 1912–1945*
*General Electric Research Laboratory*

# Preface

October 23, 1973, must have been destined to carry the General Electric monogram! By uncanny coincidence, it saw the following things happen: A Shareowners Meeting was held in Chicago. From Stockholm came the news that Dr. Ivar Giaever of the Company's Research and Development Center in Schenectady was to share the Nobel Prize for physics in 1973. At the Center, a reception was held to honor Dr. William D. Coolidge on his hundredth birthday, with Dr. Coolidge cutting a cake that bore the correct number of candles. In Washington, the Smithsonian Institution held a commemorative program marked by a similarly decorated cake.

An address given as part of that program has been combined with another delivered one day earlier at the Center to make Chapter 1 of this little book. The chapter contains an enclave called "Finishing Touches" that merits the reader's attention. The enclave is a record, largely by Dr. Coolidge himself, of his distinguished career in the General Electric Company. As he takes us through his trials, tribulations, and triumphs—triumphs that contributed much to the noteworthy success of the young Research Laboratory, he tells us a great deal of himself.

Dr. Coolidge and ductile tungsten are forever joined. The metal is worthy of the man. Tungsten is a metal of such distinction that it leads even the Britannica* to rhapsodize a little in saying that tungsten "excels all other metals" because it has the highest melting point, the highest tensile strength, the highest musical note, and the lowest coefficient of expansion; and because it can be drawn to make the finest wire. Some would argue that "highest boiling point" is another superlative that belongs in the list, which is impressive enough to justify another "Of thee I sing, baby!" even as it stands. When Dr. Coolidge met tungsten, its wire was as brittle as glass wool; when he left it, such wire could be threaded through a needle and used for sewing on buttons. Dr. Coolidge improved an already exceptional element by making it strong and ductile. He thus made it possible to replace the fragile tungsten-filament lamp invented abroad with a series of improved lamps that culminates in the admirable incandescent light sources we take for granted today.

Usefulness for ductile tungsten did not come easily, nor were the troubles over when the Research Laboratory had done its work. The public has too often been led to believe—TV commercials *not* to the contrary—that some laboratory has just brought forth another wonder, and that the millenium is finally at hand. "It ain't necessarily so!" Chapter 2 hints at what can happen to a new prod-

*Encyclopedia Britannica*, **22**, 316 (1971).

uct on its way to the market. The chapter is in the main a tale of two cities, Schenectady and Cleveland, joined to a record of battles in court that show our patent system in an unusual light.

Who should read the book? Why, all who ought to know how research, primarily industrial research, affects our civilization. Who will read the book? Let us hope, more than a few. To smooth the road for those who do, plain words have been favored where possible over technical jargon. Some of the material has been used previously to establish background in a seminar at Texas A&M University on research management. The author got to know the General Electric Company reasonably well during thirty-odd years, but he has tried to resist its influence and to write an objective book. How well he has succeeded is for the reader to judge.

Now, a personal note. During World War II a large railroad with energy on its mind (yes, even then!) was trying to suspend coal in oil. A quick method was needed to establish how fast the coal settled out. Dr. Coolidge asked me to apply x-ray absorption to the problem, by measuring x-ray intensity with a phosphor-photoelectric detector developed by H. M. Smith for the automated testing of proximity fuses. Dr. Coolidge pointed out that his proposal would probably fail because carbon and hydrogen absorb x-rays almost alike. "But," he added characteristically, "try it anyway." The sulfur in the coal unexpectedly made the method feasible. What Dr. Coolidge told the railroad, I do not know. I do know that he gave my colleagues and me the entree to a research activity that profitably filled our spare time for many years. This book is a token repayment.

So short a book and so many to thank! First, Dr. Audrey B. Davis, Curator, Division of Medical Sciences, Smithsonian Institution, for arranging the celebration in Washington of Dr. Coolidge's hundredth birthday. Next, from the General Electric Company, past and present: Dr. A. M. Bueche, for authorizing publication of the book; Messrs. R. Ned Landon and George Wise, for making the typescript into a book; Dr. C. G. Suits, Dr. A. E. Newkirk, Dr. J. H. Keeler, Dr. L. V. McCarty, Mr. Merrill Brown, and Mr. V. D. Manti, for comments, suggestions, and other valuable help. Then, my wife, Sybil Small Liebhafsky, for typing and editing; and my son, Douglas S. Liebhafsky, for unearthing from the legal literature the judicial opinions in the patent suits. Finally, three organizations. The American Society of Metals, Metals Park, Ohio, for making available proofs of W. D. Mogerman, *Zay Jeffries*, which should be in print when this book appears. Without the Jeffries biography, the Cleveland contribution to the history of ductile tungsten would have been more difficult to tell, and Chapter 2 would have suffered. The Robert A. Welch Foundation, Houston, Texas, and the General Electric Company for bearing certain expenses incidental to the book.

I alone am responsible for opinions in the book and for its shortcomings.

HERMAN A. LIEBHAFSKY

*College Station, Texas*
*May, 1974*

# Contents

# Plates

# WILLIAM DAVID COOLIDGE

Over a period of *sixty years*, ductile tungsten and the Coolidge tube have brought Dr. Coolidge international recognition too extensive for detailed description. The recognition has come from governments; from universities; and from organizations devoted to widely different fields such as science, engineering, mining, medicine, dentistry, and (of course) x-rays. The story continues: Dr. Coolidge is still being honored in his hundredth year.

Whitney, Coolidge, and Langmuir[8] are the three scientists most responsible for the brilliant success of the Research Laboratory in its first quarter-century. They complemented each other, and they interacted so effectively as to provide the best possible evidence for Whitney's firm policy of cooperation among the staff. It is difficult to segregate their individual contributions.

In the lobby of the Research and Development Center at Schenectady, there are on display fifty-odd medals awarded to its staff and to that of its precursor. It is significant that forty-odd of these medals were awarded to Whitney, Coolidge, and Langmuir, all of whom began work in the Laboratory in 1909 or before. Other able men are known to have been in the Laboratory before 1909, and many more have joined since. Why the recognitional imbalance?

I come now to a most welcome digression. It was announced this morning (October 23, 1973) that Dr. Ivar Giaever of the Research and Development Center will share this year's Nobel Prize in physics with Dr. Leo Esaki of IBM and Dr. Brian D. Josephson of Cambridge University. I am sure Dr. Coolidge feels that there could have been no better way of observing his hundredth birthday.

Let us return now to reasons for the recognitional imbalance that Dr. Giaever has so happily reduced. Justice

Holmes, whom I trust Washington has not forgotten, said of a great man as exemplified by John Marshall that ". . . part of his greatness consists in being *there*."[9] Whitney was *there* when those in control of a young growth industry decided the time had come to found the first industrial research laboratory. Coolidge was *there* when Whitney was authorized to recruit a scientist to work on the lamp-filament problem. I can no more separate Dr. Coolidge from these fortunate circumstances than Justice Holmes could separate Marshall from those that made Marshall Chief Justice. But *thereness* is far from enough: a man must by nature and training be *bound to rise*.

Would the recognitional imbalance have been as marked had Dr. Coolidge not discovered ductile tungsten? I think not. This discovery and its consequences were bound to bring great and continuing recognition to the key men involved, and these were Dr. Coolidge as prime mover, Dr. Whitney as Director, and Dr. Langmuir as we shall see.

Ductile tungsten helped bring Langmuir into the Research Laboratory. His work with Nernst had been on chemical reactions under conditions in incandescent lamps. During the fall of 1908, Langmuir met Fink, a former classmate at Columbia University, at a scientific meeting in Schenectady, whereupon Langmuir entered the Research Laboratory for summer employment in 1909.[10] He soon appreciated that ductile tungsten was better for the kind of research he had done with Nernst than anything that had been available to him in Germany, and that the Research Laboratory possessed unrivaled equipment for producing high vacuum.[11] Upon being made an offer by Whitney, Langmuir[10] terminated his "stultifying

Plate 1. Laurence A. Hawkins (1877–1958), to whose memory this book is dedicated. A man of many talents and a man for all seasons, whose contributions to research the scientific literature does not record.

stretch at Stevens Tech"; joined the Laboratory; began by continuing his German work under better conditions; applied his pure-research approach to the problem of the tungsten-filament incandescent lamp; and went on to make scientific and industrial history and to receive the first Nobel Prize awarded to a General Electric employee.

Langmuir soon began to investigate electron emission in vacuum, another kind of research for which ductile tungsten was indispensable. In 1913, Langmuir and Rogers occasionally found unexpectedly high electron emission from lamp-filament tungsten that had been doped with thoria as a consequence of the Battersea incident. After World War I, Langmuir[12] brilliantly traced this beneficial result to the presence of a sometimes partial, sometimes complete, monolayer of thorium on the tungsten. The Battersea incident thus had an equally spectacular sequel. We all know about the importance of chance in research, but this seems almost too much—even for science fiction. Let me recapitulate as best I can.

1. The Research Laboratory buys fire-clay crucibles made at Battersea, among others.

2. During the heating of tungstic oxide, these Battersea crucibles transfer something that inhibits grain growth in finished tungsten when no other crucible had previously done so. That the transfer can be effective is itself a miracle, for the finished tungsten is a long way down the road from tungstic oxide.

3. The oxide of thorium is added to tungsten powder as a replacement for this something.

4. Thorium just happens to be the best of all elements for increasing the usefulness of tungsten as an electron emitter.

5. By proper heat treatment, the *thoria* in the lamp filament can be made to yield the persistent monolayer of *thorium* needed to achieve increased emission.

Could anyone have predicted this chain of events? Remember, the chain could have been broken at any link. I think Dr. Coolidge would point out, politely but firmly, that significant research results are not yet predictable by computer or otherwise, and that experiments have not yet had their day!

## THE DIRECTOR OF RESEARCH

In 1932, Dr. Whitney chose to retire as Director of Research, in part because his health had begun to suffer owing to strains generated by the depression of 1929, which had begun to affect the Company seriously by 1931. The number of employees was reduced; the workweek was shortened; and additional reductions in take-home pay had to be made. The Research Laboratory was not immune.[13] There was talk during the early thirties that the Laboratory would be closed.[14] These were the conditions under which Dr. Coolidge succeeded to the Directorship, a position that continues to demand—even under the best of circumstances—all that a man can give. Dr. Coolidge was equal to the burden he suddenly had to assume.

When Dr. Coolidge took charge of General Electric's "Adventure into the Unknown," the adventure had already succeeded. The success began with the discovery of ductile tungsten, which not only made possible an expanded lamp business but initiated an *electronics era* that is still with us, though altered in form owing to the introduction of solid-state devices. By 1910, the *materials* ductile and thoriated tungsten had become available as a result of Dr. Coolidge's work. Materials usually attain maximum usefulness only when there is an understanding of the *processes* of which they are capable and to which they are subject. The need for such understanding often replaces necessity as the mother of invention. Between 1910 and 1930, understanding of electronic processes and invention of electronic devices progressed together. The thoriated tungsten emitter (Langmuir) and the Coolidge tube are but two examples. Others impossible to omit are the understanding of space charge by Langmuir,[27] and the shield grid tube and the Thyratron, both invented by A. W. Hull. Of course, many electronic devices do not contain tungsten; but tungsten was a major factor in making the electronics era even by 1930 so successful that it had become a hard act for any research laboratory to follow.

In June 1933, within a year of taking charge, Dr. Coolidge made an important move when he asked A. L. Marshall to head the Insulation Section of the Laboratory. In the letter of appointment,[15] Dr. Coolidge wrote: "We may later want to broaden the title, as the field of activity which I have in mind includes such other chemical activities as those which you have been carrying on in the past." Dr. Coolidge did more. Even before the Company had recovered from the depression, he made it possible for Marshall to undertake a vigorous program aimed at a considerable expansion of what had become the Chemical Section. The expanded Section responded in part with four new and important polymeric materials—Flamenol, Formex, the Permafils, and silicones—before the end of World

War II. Anyone who knows the value of new materials to the General Electric Company would agree that a polymer era accompanies the electronics era in the history of the Research Laboratory.

Dr. Coolidge moved early and decisively as regards nuclear energy.[16] The field was opened early in 1939 by Hahn and Strassman, and by Meitner and Frisch. On May 23, 1939, Langmuir wrote in his notebook that this "recent discovery . . . makes it very important (now for the first time) to . . . work in nuclear physics and particularly to develop methods of separating isotopes on a large scale. . . . I talked all this over with Coolidge yesterday. . . ." By May 24, Dr. Coolidge had asked K. H. Kingdon "to look into this," and he had been "digging up the literature and discussing it with [H. C.] Pollock and others." After subsequent discussions, Kingdon and Pollock in early June had visited two universities to inquire about three methods of isotope separation. By June 12, Dr. Coolidge had authorized them to proceed with an investigation that still ranks high as an example of what could be accomplished in less than a year[17] by Research Laboratory personnel with Research Laboratory facilities in an important new field where everything had to be done from scratch; including, for example, the design and the building of a mass spectrometer capable of separating ($U^{235}$ + $U^{234}$) from $U^{238}$ in amounts large enough for significant cyclotron experiments. In October 1940, Dr. Coolidge received from Dr. Lyman J. Briggs, Director, National Bureau of Standards, an order for 150 grams of $UF^6$, then a large amount, which was needed for various isotope-separation projects. The order was quickly filled. But military necessity forced the government to preempt the field in November 1940, and the Research

Plate 2.   Dr. Coolidge with the nucleus of his ductile-tungsten task force at the Research Laboratory in 1911. From the left: unidentified, Leonard Dempster, Johnson Hendry, George Hotaling, and Dr. Coolidge.

14

Laboratory lost an opportunity to do work that might have been as useful to the Company as the research on tungsten.

In 1940, Dr. Coolidge, then in his sixty-seventh year, appointed Dr. C. G. Suits to the specially created post of Assistant to the Director, an action interpreted as indicating that Dr. Coolidge planned to retire and that Dr. Suits would be his successor. World War II intervened.

With the Research Laboratory's record in World War I, it was certain that the Research Laboratory under Dr. Coolidge would cooperate fully with the government in World War II. As the military uses of research were more fully understood when the second war approached, virtually the entire Laboratory came to have at least some connection with government-contract work.[18] Throughout the war, Dr. Coolidge served as consultant to the National Inventor's Council and the National Defense Research Committee while he continued to direct the Laboratory with the help, now more nearly indispensable than ever, of L. A. Hawkins, Executive Engineer. Dr. C. G. Suits meanwhile became Head of Division 15, NDRC, radar countermeasures, the field of the Laboratory's largest war effort. On January 1, 1945, Dr. Suits succeeded Dr. Coolidge, whose activity in and outside the Company[19] did not end until long after his retirement at age 71.

"At its beginning, the Research Laboratory was Whitney, and Whitney was the Research Laboratory. In one very real sense that has been true throughout the laboratory's history. . . ." This appraisal,[20] made by Hawkins in 1950, is applicable to 1932, when Dr. Coolidge became Director, but it was bound to becomes less applicable with the passage of time.

A large industrial research laboratory changes subtly and continuously, during the intervals between easily recognizable abrupt changes such as reorganization, and changes attributable to depressions or wars. In judging how a laboratory is managed, one must remember that "it is not possible to step twice into the same river."[21] Perhaps Dr. Coolidge diverged most markedly from Dr. Whitney's major policies by moving toward formal organization and increased decentralization, concerning which he said in 1937:[22] ". . . every large laboratory of which I know is organized on the group system. That becomes practically essential with increasing size. There is a limit to the number of workers one man can supervise. . . ."

This article contains a superb description of the Whitney-Coolidge conception of the proper role in 1937 of a research laboratory in a large industry. Dr. Coolidge says:[22]

> I do not wish to create the impression that all our work, or even most of it, is of a fundamental nature. It actually engages only perhaps twenty per cent of our activities. Our laboratory is truly a *service department*, even though the service we render is only advisory. Problems are continually arising in other departments in which our help is needed. As far as possible we refer such problems to the various works laboratories, but many are of such a kind that we, with our highly trained specialists and special facilities, are best equipped to tackle them. In such cases, we undertake the job, and in general try to *give it precedence over our own problems*. Such service work amounts to some forty per cent of our total.
>
> Again, it frequently happens that the possibility of developing some radically new device is suggested by our researches, or by some other event, and it may be so different from any current product that no existing department has the

knowledge, experience, or facilities to undertake the development efficiently. It then is needful for us *to develop it* and perhaps *even to manufacture for a time.* Graphitized brushes, tungsten ignition contacts, X-ray tubes, radio power tubes, carboloy, and various forms of glyptal resins, were first manufactured in the laboratory. We never manufacture when another department is ready to do so, nor do we carry a development farther than the point at which another department is prepared to take it over, but nevertheless such development work constitutes another forty per cent of our total.

But it is our fundamental work [pure research] which we consider *our most important activity,* and which we jealously guard against encroachment by service and development work. It is from that that we hope will come the radically new things to add new and profitable lines to our company's product, to broaden its field of interest, to give new employment to labor, and to benefit the public through new comforts or conveniences. Moreover, it is invaluable in its effect on our whole laboratory staff. Each contribution we can make to the advance of scientific knowledge increases our contacts with other workers in science, through attendance by our men at meetings of scientific societies for the presentation of papers, and through interchange of visits, induced by similarity of interest and of effort. I now recall with amusement my fear, thirty-two years ago, of being pocketed and isolated at Schenectady. [Emphasis and bracketed material supplied.]

Dr. Coolidge was probably ahead of his time in viewing the Research Laboratory as a services component; such components were formally named in the Company only after World War II. Of course, the services rendered by the Laboratory today are different from what they were in 1937—to vary the earlier metaphor, much water has by now gone over any dam in the river of Heraclitus.[21]

W. D. COOLIDGE.
TUNGSTEN AND METHOD OF MAKING THE SAME FOR USE AS FILAMENTS OF INCANDESCENT
ELECTRIC LAMPS AND FOR OTHER PURPOSES.
APPLICATION FILED JUNE 19, 1912.

1,082,933.                                    Patented Dec. 30, 1913.

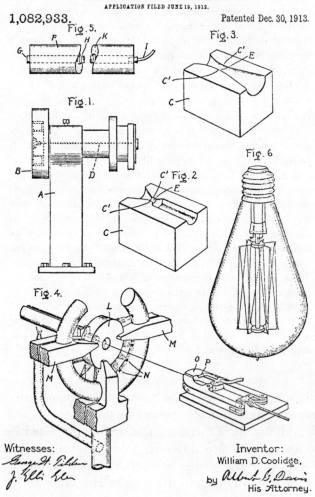

Witnesses:

*George H. Pilldern*
*J. Ellis Glen*

Inventor:
William D. Coolidge,

by *Albert G. Davis*
His Attorney.

## FINISHING TOUCHES

It is not often that a great scientist makes a concise summary of his career. Dr. Coolidge did just that when he prepared a digest of his many laboratory notebooks. Items from this digest, identified by their dates and regrouped without significant changes, are used below as Dr. Coolidge's own finishing touches to the preceding sketch of him and to the outline of his work. These items are supplemented by quotations from his articles and by the present author's comments to provide continuity.

### Ductile Tungsten

*2/19/06.* Tungsten work: pressed-powder rod melted in improved design of Weintraub mercury-arc furnace. Hot

---

Plate 3. The cover sheet of the Coolidge patent on ductile tungsten. Figure 1, a swaging machine. Figure 2, one form of swaging die. Figure 3, a second form of swaging die. Figure 4, one form of wire drawing apparatus. Figure 5, an electric furnace devised by H. A. Winne and G. Dantsizen, two other notable General Electric men. Figure 6, an incandescent lamp with straight-line tungsten filament (now obsolete).

In Figure 1 B is a rotating head that brings together pairs of dies like those in Figure 2 or 3 to reduce the cross-section of the tungsten, which enters at the left and leaves through the passage D. The dies, invented by Dr. Coolidge, differ from the conventional in the shortness of their faces (E) and in having rounded edges (C') to keep the die from digging into the tungsten rod. For the drawing of wire as in Figure 4 the die (L) is a drilled diamond set in metal, O is a clamp to hold the pointed wire first put through the die, and P is a gas jet to supply heat.

hammering of melted-tungsten globules and of heated ends of the pressed rods. Some hot malleability.

COMMENT.   When Dr. Coolidge began trying to make tungsten ductile, it was known to be soft (hence ductile) at the operating temperature (say, 2500° C) of an efficient incandescent lamp, but there was, no way of working the metal at this temperature or even at a temperature very much lower. The prevailing belief was that tungsten at room temperature was *inherently* brittle. How did the problem look in 1910, some years later?

> To a man ignorant of our success, the problem would certainly look more hopeless today than it did [at the beginning]. For since that time millions of tungsten filaments have been produced from all available tungsten ores, by widely differing methods, and by different groups of men. And each manufacturer has been fully alive to the fact that he must strive for the highest attainable purity. Yet all of the filaments made have been brittle. They are elastic and flexible as spun glass, but, like the latter, are incapable of taking the slightest permanent set.[23]

* * *

*11/26/06.*   Palmer taught us how to make diamond dies; or rather how to drill diamonds. He was here 27 days. His bill for services was $300 [sic!].

*2/21/07.*   Successful hot-rolling of tungsten filaments (10.5 mils in diameter) between heated rolls.

*10/21/08.*   Fink showed me a piece of tungsten which bent cold. It was a 9.8-mil filament that had been drawn through 5 dies.

COMMENT. The first entries relate to squirted filaments, which became obsolete when ductile tungsten resulted from the hint of 10/21/08.

*　　*　　*

*12/14/08.* With C. A. Cowles of Ansonia [Ansonia
Copper and Brass Company], studying drawing, swaging, and
rolling of copper wire. We perhaps should have a swaging
machine.

COMMENT. Swaging a sintered bar to form a rod is the
first step in the working of tungsten to make drawn wire.
A swager is a complex rotary hammering machine that
must meet severe requirements if the processing of tungsten is to succeed. Dr. Coolidge had to invent new swaging methods that flew in the face of all previous experi
ence. See Ref. 5a, pp. 100–106. Swaging was followed by
drawing the resulting thin rods through dies to make wire.

*4/1/09.* Now have a swager.

*7/16/09.* Swaged tungsten rod (⅜'' square) down to 128
mils; then to 78 mils; then to 53.5 mils—then bent it
cold.

*　　*　　*

*9/11/09.* Before leaving for vacation, asked Barlow and
Hendry to make many tungsten rods for Hotaling to use
on return from vacation. They were sintered in porcelain-
tube furnace—all proved hopelessly bad, breaking in first
die. Trouble, oxygen inside—tungsten is too fine. [The
men named worked with Dr. Coolidge.]

*1/5/10.* It's now clear we can't use fine tungsten powder [to make wire] because we can't get oxygen out.

COMMENT. This led to the heating of tungstic oxide needed to give particles that could be reduced to form coarse tungsten powder—next stop, the Battersea incident!

*1/12/10.* Non-offsetting tungsten from crucible-fired oxide. [The Battersea incident!]

*6/21/10.* Before leaving Schenectady, I asked Hotaling to add erbium and yttrium oxides to tungsten made from unfired [no Battersea treatment!] oxide. I find he got very fine grains: non-offsetting structure in both sintered and swaged rods. I am also trying $ZrO_2$ and $CeO_2$. All these rods give off much smoke at the high treating temperatures and show fine grains. [Note awareness that oxides inhibited grain growth.]

*9/27/10.* Reeling tungsten wire, to handle long lengths.

COMMENT. A triumphant close to what I believe is the most important applied research project yet carried out in the General Electric Company.

The work which has been outlined above is the result of the close cooperation of about 20 trained research chemists, with a large body of assistants, in the research laboratory. These men were of course given, from the factory organization, all of the mechanical and electrical assistance they could use, and were assisted in no small measure by the staff of the incandescent lamp factory. [From p. 965, Ref. 23. Noteworthy as an early example of cooperation between research and operating components.]

Plate 4. During a visit by Thomas Edison to the Research Laboratory in 1914 Dr. Coolidge, with needed emphasis, explains how tungsten is made ductile by use of the apparatus in the foreground. The tungsten enters the furnace in front, is heated to swaging temperature, and then reduced in cross-section by the swager, rotated by the electric motor.

COMMENT. By 1912, Dr. Coolidge could report: "This development [of ductile tungsten] has taken place so rapidly that today [1912] the bulk of the world's supply of tungsten lamps is made from drawn wire."[24] The world seems to have been hungrier for ductile tungsten than for the proverbial mousetrap.

### The Coolidge Tube

COMMENT. X-rays were discovered by Roentgen on November 8, 1895. Probably no earlier scientific discovery had impressed the world so quickly and deeply. Dr. Coolidge did not escape. At home the following summer, when he had graduated from MIT, he built a large electrostatic machine to produce high voltages for the generation of x-rays. For the building of it, he had the use of a machine shop in Hudson, Massachusetts, owned by the father of a friend. Young Coolidge sold the machine to a local doctor for x-ray work.[25, 26]

X-rays never left Dr. Coolidge's mind, as the next two items show. Although fully occupied with the lamp-filament problem, he was looking ahead to the time, years distant, when he could begin research on x-ray tubes.

*2/12/06.* Tungsten or tantalum could be used to advantage in x-ray tube because of higher molecular weight. Tungsten better because of higher density (atomic number).

*4/8/09.* Uses for ductile tungsten and molybdenum. Projectiles, x-ray targets, contact points.

\*\*\*

*5/8/12.* Received from Wiggins first tube with solid tungsten cathode and a heavy tungsten anode without backing.

This is the crankiest tube I've seen. Cathode got hot, and I found that with cathode bright red, the tube showed entirely different characteristics. Vacuum became stable and stays so as long as cathode is bright red. Tube seems to show less local heating of glass around cathode and on front hemisphere than regular tubes. Perfect steadiness is remarkable. Cathode heating is localized at center.

COMMENT. The usual x-ray tubes at this time contained a gas at low pressure, the positive ions from which bombarded the cathode to generate electrons that bombarded the anode to generate x-rays, charged particles of both kinds being accelerated by the field. The "entirely different characteristics" mentioned above probably arose from better-than-usual vacuum, which would make this a first approach to the Coolidge tube.

\*\*\*

*12/10/12.* Irving Langmuir tells me that in his study of the Edison effect, current from hot cathode is greater with vacuum of .01 or .02 micron than at higher pressure (except in case of argon). I will try this at once in an x-ray tube in which I can heat cathode to any temperature.

COMMENT. The Edison effect is an electrical discharge seen by him in incandescent lamps. Langmuir's highly important work on this effect in tungsten-filament lamps led to an understanding of the influence of space charge and of residual gases on thermionic emission in high vacuum—hence to an early understanding of vacuum tubes.[27] The last sentence of Dr. Coolidge's notebook entry shows that he saw immediately the importance to the x-ray tube of having the electron current controlled only by the temperature of the filament: this meant that current and voltage would be *independently* variable. The work in this direction was brilliantly successful. Gas (or ion) x-ray tubes had been temperamental, short-lived, hazardous to use, and unreliable. The Coolidge tube soon became an indispensable and satisfactory, self-rectifying, precision instrument.

\* \* \*

The following quotations from the paper[28] in which the Coolidge tube was announced tell much about the character of its inventor:

> The idea of using a hot cathode in a Roentgen tube was not new, but, so far as the writer could learn, the principle had never been successfully applied in a vacuum good enough so that the positive ions did not play an essential role.
>
> The work of Dr. Langmuir has shown that a hot tungsten cathode in a very high vacuum could be made to continuously yield a supply of electrons at a rate determined by the temperature. [Thus Langmuir repaid Coolidge for ductile tungsten.]
>
> The result of efforts in this direction has been entirely suc-

cessful and tubes have been made, based upon this principle, which are free from all of the above-mentioned limitations [of gas tubes].

\* \* \*

*1/21/14.* Noticed yesterday hair coming out from back of head and scalp. Sore there, due to 2 Frontal Sinuses [x-ray exposures] made [by Dr. Coolidge] 12/5/13.

\* \* \*

*4/5/17.* Thomas Cooke here yesterday re development motor-truck [x-ray] outfit. He's on train with me going to New York. The committee on field-service outfits [military x-ray equipment] meets tonight in New York. The Delcolight outfit [portable power source developed with help of C. F. Kettering, "Kett," then Delco Company, later General Motors] is now in excellent shape.

*5/9/17.* Trying Delcolight Outfit No. 2. Kett is sending man to put in other field coils.

*6/14/18.* Am interested in light-weight [x-ray] outfit for doctor to take in auto to bedside of his patient. On 60-cycle City Service. [Will] need no rotary [equipment to supply needed voltage].

\* \* \*

COMMENT. Dr. Coolidge continued applied research on x-ray tubes until some years after he became Director in 1932. The portable sets for service at the front led logically to the dental x-ray unit, in which a small, self-rectifying, self-shielding Coolidge tube was wedded to a 56-kilovolt supply transformer enclosed in an oil-filled case, the unit being small and light enough to be mounted on a swinging bracket for convenient use on a patient in the dentist's chair. The medical profession, which quickly became enthusiastic about the first Coolidge tubes as diagnostic aids, soon wanted x-ray tubes for deep therapy. For this use, the energy of the x-rays, which governs their penetrating power, had to be increased. This brought two major problems, both of which Dr. Coolidge solved. The first was the need to improve heat removal from the x-ray target. This was done by flowing cooling water through the target. The second was the need to increase tube voltage up to 1000 kilovolts and more. Dr. Coolidge did this by inventing and applying the "cascade principle," according to which the effect of tube voltage is "cascaded" by dividing the voltage over a series of tube sections: each section is independent as regards voltage gradients, but each in its turn accelerates the electrons that eventually bombard the target (or anode) to produce x-rays. Dr. E. E. Charlton and Mr. W. F. Westendorp were Dr. Coolidge's principal colleagues in this work.[29]

Although the primary push for new types of Coolidge tubes came from the medical profession, these admirable x-ray sources soon found increasing use in analytical chemistry and in nondestructive testing. See Ref. 4 and its successor.[30]

The importance of ductile tungsten in vacuum tubes (e.g., tubes with thoriated tungsten electron emitters) and in other applications will have to be taken for granted.

Plate 5. Panoramic view of the most important section in the General Electric Wire Plant, Cleveland, Ohio. A fitting sequel to the preceding plate! The tungsten wire-drawing area processes wire from heavier sizes at the rear to very fine sizes (0.0005 inch in diameter) in the foreground. These hundreds of machines produce wire that is then sent to the many lamp plants where the wire is wound into filaments and assembled into lamps.

## Cathode-ray Tubes

*2/14/14.*   As soon as possible, want to get *very penetrating* cathode rays out through metal window.

COMMENT. The key words are italicized above. Lenard had done similar work with cathode-ray (electron) beams, and had helped Roentgen to start in October 1895 the cathode-ray work that led him to the discovery of x-rays. In principle, cathode-ray and Coolidge tubes are close relatives. In the former, electrons are taken out of the tube through a thin metal window. In the latter, the electrons strike a metal target, and the x-rays thus generated are taken out through a special window, usually of beryllium. Dr. Coolidge's work on cathode-ray tubes was successful and led to interesting experiments, among which only the following—too good to omit—will be included.

*11/14/24.*   Wonderful results with latest cathode-ray tubes on 240,000 volts—glow, 8-inch radius calcium oxide gives brilliant red fluorescence—glass exposed to beams, as much darkened in 5 seconds as in 3½ hours of x-ray [exposure] at 14 inches with 30 milliamperes and 200,000 volts. Asked Tanis to go ahead using molybdenum-window support that Hotaling has made. He is to try first a 0.5-mil aluminum window. If OK, then a 0.3-mil molybdenum-window fastened on with copper. [Planning of multi-sectional cathode-ray tubes for higher voltages began during the following week.]

\*\*\*

## Other Research

COMMENT. In accord with the Whitney-Coolidge research philosophy discussed earlier, Dr. Coolidge concerned himself with many problems unrelated to his principal activities. Examples follow: submarine detection, World War I; high-power Kenotrons (vacuum tubes), 2/10/22; high-voltage cable, 5/28/23; power rectifier for electric locomotive, 11/5/23; quieting of domestic refrigerators, 12/17/25; 9/13/27; 1/17/28; vacuum switches, 3/25/27.

* * *

## Personal Contacts and Travel

From the first, the Research Laboratory recognized the advantages of extensive contacts at home and abroad; for this liberal policy, the critical incandescent-lamp situation before the invention of ductile tungsten was partly responsible. Dr. Coolidge has traveled much, both before and after his retirement; for example, around the world in 1959, near age 86. His visitors, noted chemists, physicists, and physicians among them, came to be so numerous that they must have interfered with his work. His comment: "In the long run, we in the laboratory have benefited as much from our visitors as they have benefited from us."[31]

* * *

*9/2/08.*    Back from German trip [mainly to discuss tungsten filaments. Apparently, no significant results.][32]

*10/6/09-12/3/09.*    Trip abroad. [This time, Dr. Coolidge carried samples of the new ductile-tungsten wire with him.]

*10/16/09.*    With Remané, Blau, and Kallman, forenoon. [Dr. Fritz Blau, chemist, was a high-ranking employee of Deutsche Gasglühlicht Aktiengesellschaft and was interested mainly in filaments for incandescent lamps.]

*11/5/09.*    Had appointment with Blau, forenoon, which he postponed until afternoon.

*11/6/09.*    Cabled Mr. Morrison regarding exchange of information about [ductile-tungsten] wire. [Mr. Morrison was an important patent counsel in the General Electric Company.]

*11/8/09.*    Blau reluctantly consented to appointment for afternoon. [WDC] made full disclosure regarding drawn tungsten and molybdenum then.

COMMENT.    When Dr. Blau became convinced that he was actually holding a sample of ductile-tungsten filament wire, his reaction was understandably violent; he realized that this slender thread portended a revolution, bound to be painful for everyone but General Electric, in the lamp industry; and that American industrial research had scored a signal victory (probably its first major victory) over European. To be continued in Chapter 2!

*7/26/20.*    At noon, office [in Paris] of Pilon and Gallot [electrical manufacturers]. Luncheon my honor, by CFTH [the French Thomson-Houston Company, another electrical manufacturer], and Pilon and Gallot at the Club de la Renaissance: Saurel, Pilon, Gallot, d'Arsonval, Abraham, Duc de Broglie, Dauvillier, Haret, Henri Beclerc, L. L. Blondin, Cahen, and Gunther. I gave short talk. [A long way from Coolidgeville near Hudson, Massachusetts! Typical of the recognition that Dr. Coolidge received abroad after ductile tungsten and the Coolidge tube became well known.]

## General Electric Career

*9/26/05.*  First entry in first notebook refers to breakage of tantalum filaments on alternating-current operation. [Trouble came early!]

*11/24/08.*  Raise of $1000 beginning 3 months from today.

COMMENT.  "Coolidge arrived to commence work"— notebook entry by Dr. Whitney, September 11, 1905.[33] Dr. Coolidge's counterproposal to Dr. Whitney's offer had been: twice the MIT salary and half-time on his own research with needed equipment supplied by the Company[34] [No matter how it looked at the time, it was a bargain!]

* * *

*6/19/28.*  Tuesday. Ass't Director from now. Start Laboratory Council meeting every week. Whitney, Coolidge, and Langmuir.

8/22/28. Wednesday. As member of committee of four, have been visiting plane and engine factories [not GE] since July 31.

\*\*\*

2/6/30. Have been ill for 2 or 3 days. Whitney has arranged for me to assume full responsibility for Kimball, Newkirk, Taylor, Hewlett, and their men. This includes responsibility for fixing salaries even. [A. L.] Marshall [later, Head, Chemical Section] leaves Saturday for South, so I must look after his work.

\*\*\*

10/21/32. Whitney told me yesterday morning he's asked Mr. Swope [then chief executive officer of GE] to make me Director, to take office November 1, 1932.

11/2/32. Became Director yesterday.

## The Time Clock

4/13/33. Burrows and Muir both opposed to time clocks for section heads.

COMMENT. At this time, every—repeat, every—General Electric employee in Schenectady "punched the clock." Dr. Coolidge was determined to get this changed in the Research Laboratory. Mr. Swope was not immediately responsive.[35] Dr. Coolidge must have begun his campaign on or before the date given above. Mr. Burrows (manufacturing) and Mr. Muir (engineering) were two of the most important men in the Schenectady Works.

In 1936, the liberation of research personnel from clock-punching began with the very few people whose salaries were $6000 or more. The liberation eventually became complete. [HAL, personal knowledge.]

* * *

### The Experimental Approach

"We may define research as the systematic search *by the experimental method* for new knowledge . . ." [Ref. 22, p. 10. Emphasis supplied.]

"The successful outcome of our ductile tungsten work was due to the right microstructure. I must, however, say that we were guided, in the main, by the experiment itself rather than by metallurgical knowledge."[36]

* * *

Whitney wished to demonstrate the ionization of air to a large scientific audience. An assistant had failed in a week to find a satisfactory method. "Whitney told Coolidge of this at lunch one day. He had not been back in his office a half hour when Coolidge called him to one of his workrooms. There were three toy balloons hung by long strings from a single support, so that they rested in contact with each other, and a portable x-ray outfit pointed

Plate 6.   Dr. Irving Langmuir (*left*), Sir Joseph J. Thomson, the famous British physicist, and Dr. Coolidge are shown in 1923 inspecting a pliotron, one of the high-vacuum tubes invented at the Research Laboratory near the beginning of the electronics era.

in their direction. Coolidge electrified the balloons by rub-
bing them on his hair. The like charges repelling each other
then held the balloons two feet or more apart, but almost the instant the switch was closed on the x-ray
outfit, the balloons dropped into contact again. The rays
had ionized the air around the balloons and made it con-
ducting, so that their charges almost at once leaked
off."[37] [Note the quick insight; the simple, elegant experiment; the taking of immediate action; and the verifica-
tion before disclosure—no wonder Hawkins considered Dr.
Coolidge to be the Laboratory's most resourceful
experimenter![38]]

\* \* \*

### Further Personal Traits

COMMENT.   Dr. Coolidge's many admirable personal
traits are easily inferred from what has preceded and from
the items to follow. To spell out these traits is not necessary and would, I believe, embarrass him. I do ask you to
remember his intense *dedication* to science and to work; his
*toughness,* mental and physical; and his *integrity*—additional
evidence for all of which will now appear. "Toughness" as
used here has no pejorative connotation. I have never
known a more polite, tactful, and considerate man than
Dr. Coolidge.

5/12/09.   Visited C. A. Cowles in Ansonia. He was willing to lend me his large diamond die for drawing molybden-

um, but I was afraid. [Prudence, not fear. See item dated 12/14/08 under Ductile Tungsten.]

*8/12/09.* Hydrogen explosion in tube furnace blew flame and sand into my eye. 1 visit to hospital and 7 to [doctor's] office. $40, *which I paid.* [Emphasis added.]

*8/1/10.* [Name deleted] threatened me—he's leaving Laboratory at my order.

\* \* \*

*12/15/26.* Received notification of award of Edison medal.

12/15/27. Received notice of award of Edison medal.

COMMENT. The medal was reawarded in 1927 because Dr. Coolidge felt bound to decline the first award, made in part for the "origination of ductile tungsten." Here is what happened. Shortly before the first award was to be made, Dr. Coolidge's discovery of ductility in tungsten was declared unpatentable in a United States Circuit Court on the grounds that ductility was a property inherent in metallic tungsten. Dr. Coolidge declined the award even though it had been made in part for "the improvement of the x-ray tube." When the citation was reformulated (see Appendix 1), Dr. Coolidge accepted the medal.[39] To be continued in Chapter 2.

\* \* \*

The Research Laboratory began research on silicones after phenyl silicones had been shown to A. L. Marshall and W. I.

Patnode on an invited visit to the Corning Glass Company. Dr. Coolidge as Director insisted that this debt to Corning be acknowledged in the first General Electric publication on silicones.[40] [HAL, from a reliable source.]

\* \* \*

˙On a guided tour through a competitor's laboratory, Dr. Coolidge was told by the guide that he might profit from looking into a laboratory where proprietary work was being done. Dr. Coolidge stiffened, quickened his pace, looked straight ahead as the door of that laboratory was approached and passed, and said no more to the guide. [HAL, from a reliable source.]

\* \* \*

In response to the suggestion that a spectrograph be bought to satisfy a competent scientist, Dr. Coolidge said, "Remember, we aren't here very long; we must think of what's best for the Company." [HAL, personal knowledge.]

\* \* \*

After a meeting during World War II, during which an Army officer with a suggestion for military research was brilliantly but tactfully cross-examined by Dr. Coolidge, Dr. Coolidge gave the notes he had made to a younger staff member (who had taken no notes!) with a polite request for a written report. [HAL, personal experience.]

\* \* \*

"But above all else is his indefatigable interest in science. After a long day in the laboratory, in which he had concentrated on a dozen different matters, which might well weary a man of more robust physique than he, Coolidge might turn to an evening of scientific discussion with a visitor, seemingly finding renewed stimulation, not fatigue, for his keen, acquisitive mind. His colleagues often wondered how he could maintain the mental pace he did, and were forced to conclude that what would be exhausting work for most men was only play or gentle exercise for him."[41]

\*\*\*

"This week was a rather hard one, as far as work goes, until yesterday. And yesterday afternoon paid for it all. I got some wire out yesterday that was ideal.

"Have spent whole day in the lab, and it's now 10:30 p.m. . . ." [Letter from Dr. Coolidge to his parents during the ductile-tungsten period, in which he sometimes worked Sundays.[42]]

\*\*\*

When a reporter asked Dr. Coolidge about the achievement of ductility in tungsten, he replied: "Don't put all the emphasis on me. In the first place, whatever I may have done, I did with the assistance of a staff of able workers, and secondly, we all had the advantages of facilities for exhaustive research provided by a financially powerful and forward-looking corporation."[43] [I do not know of any case in which Dr. Coolidge took sole credit for a worthwhile accomplishment.]

## A LESSON FOR TODAY

Let me now compare Dr. Coolidge with a great American who was active on the bench in Washington until after his ninetieth birthday. I refer of course to Justice Holmes, famous son of a famous father, both of whom I quoted earlier.

By ancestry and early environment, young Holmes belonged to the new England aristocracy defined by his father[44]—he was, in truth, a Boston Brahmin. The story of young Holmes could reasonably have been called "Already There"—certainly not "Bound to Rise"—and it would probably not have been written by Horatio Alger.

And yet, Dr. Coolidge and Justice Holmes have much of importance in common. Holmes' great sentence,[45] "The life of the law has not been logic: it has been experience," harmonizes with the experimental approach, which Dr. Coolidge values so highly. Both men have demonstrated sharp insight into complex problems—insight that has often led to simplification. Each man reached the top of his profession. Above all, the career of Justice Holmes gives evidence of the dedication, the toughness, and the integrity that characterize Dr. Coolidge.

Both men have an important lesson for us today. The institutions on which our civilization rests are so complex and so large that institutions and civilization threaten to become unmanageable. Complexity and size are interacting parameters that generate pressures resulting in unfortunate decisions and acts. Where Dr. Coolidge would ask "What is the right thing to do?", we ask today "At this point in time, have I kept all my options open, and what is the best deal I can make?" Can't we do better? Can't we simplify our civilization by showing greater dedication, toughness, and integrity? We cannot turn back the clock to the New England of the young Coolidge, nor should we wish to do so. But we need leaders for all our institutions who have, and who will practice, the best qualities of Dr. Coolidge and Justice Holmes.

On receiving the Nobel Prize, William Faulkner said: "I believe that man will not merely endure: he will prevail." To endure could mean merely to survive, deadened by commercials and deafened by rock-and-roll, in a faceless Orwellian civilization rolling on polymer tires. To prevail we must make science, technology, and government yield a more worthwhile life than ours threatens to become. Dr. Coolidge has done more than his share— now it's up to us!

# TWO

## Ductile Tungsten Leaves the Laboratory

*Many things happen*
*between the cup and the lip.*

Burton, *Anatomy of Melancholy, Part II*

Dr. Coolidge's two great contributions are ductile tungsten and the Coolidge tube. Both are classic examples of the value and the influence of industrial research at its best. As such, they are significant items in any history—certainly in any scientific history—of civilization in the twentieth century. There is no need to describe or to document their end results; they are obvious. Less likely to be noticed, but perhaps of deeper historical significance, is what happened to them after the research was done. Here ductile tungsten seems to deserve precedence over the Coolidge tube, not primarily because the first led to the second, but because of the history of its transition from research to manufacture and because of its record in the courts, both of which illustrate the intricate relationships and the growing complexity of our civilization.

## SCHENECTADY TO CLEVELAND

Industrial research costs a great deal of money. To recover the cost, a company needs transitions to operating components for its successful research projects, and these transitions must come frequently enough and be profitable enough to support also the unsuccessful research. The transition of ductile tungsten from research in Schenectady to manufacture in Cleveland illustrates what usually happens. Like the course of true love, such transitions "never did run smooth." Because names of Company components have changed, we identify our transition as a tale of two cities, although this is an oversimplification.

Dr. Coolidge virtually completed research on ductile tungsten in 1910. How did the upcoming transition look then? Very good indeed. The most important bequest the

Plate 7.   Dr. Coolidge holds an early portable x-ray tube, the forerunner of tubes such as those used by dentists today. He invented the tube in time for it to be useful in combat zones during World War I, not too long after the close of which this photograph was made.

research laboratory can make to the prospective operating component in such circumstances is the *certainty* that *something worthwhile* has been achieved *under laboratory conditions*. With ductile tungsten, the worthwhileness was so clear that no market survey was needed. A new filament wire, miles long, was at hand that made possible a less costly, more rugged, more efficient lamp of acceptable life. An important saving in energy was going to be made over the life of a satisfactory lamp that could be operated at the higher temperatures (hence higher efficiencies) attainable with filaments of ductile tungsten.[1] And the world badly needed more light at a price it could afford. The transition of ductile tungsten seemed sure to proceed with more than deliberate speed.

The General Electric Company as formed in 1892 contained the Edison Lamp Works at Harrison, New Jersey. Dr. Coolidge's operating-component contacts during the time he worked on ductile tungsten were mainly with this Works, where Mr. J. W. Howell, an engineer with whom Dr. Coolidge maintained close relations, was an important figure in the lamp industry. Ductile tungsten went first to Harrison.

In 1901, the General Electric Company organized, and in 1912 took over, the National Electric Lamp Company in Cleveland. This opened another route for the transition of ductile tungsten into operations. Two remarkable men, Mr. B. L. Benbow and Dr. Zay Jeffries, were identified with the Cleveland Wire Works, in which ductile tungsten for all General Electric incandescent lamps eventually came to be made. There are rewards for centralizing lamp manufacture. Somewhere near 100 tons of tungsten will make filaments for all the incandescent lamps the world needs in a year. A pound of tungsten, drawn into a wire 8.5 miles long, will make filaments for 23,000 60-watt lamps.[2]

Much of the information to follow comes from a biography, about to be published, of Dr. Jeffries,[3] advance proofs of which were kindly made available by the publishers for use here. The book should be read by everyone interested in tungsten (cemented tungsten carbides included) or in the career of an outstanding metallurgist who, for part of his working life, was active simultaneously in the aluminum industry, the steel industry, and the General Electric Company. Dr. Jeffries joined the Company fulltime on July 1, 1936, and retired on December 31, 1949, as Vice President and (the first) General Manager of its Chemical Division.

Between 1906 and 1911, more than $116,000—piddling now, enormous then—had been spent on ductile tungsten in the Research Laboratory.[4] Ductile tungsten wire was first drawn late in 1908; intensive and extensive work to develop a large-scale manufacturing process followed quickly. Anyone who knows Dr. Coolidge will agree that he must have believed this large-scale process to be well in hand when he left ductile tungsten to turn to x-ray work. And yet . . .

In 1914, Mr. B. L. Benbow was manager of the Cleveland Wire Works, and here are some of his early troubles: re-embrittling of tungsten wire during its manufacture; acute difficulties with swaging dies; costly breaking of dies through which wire was drawn; sagging of coiled tungsten filaments; and formation of tungsten carbide on the wire surface. According to the Jeffries biography (p. 86), Mr. Benbow was ordered at one point to abandon the Coolidge process; to scrap the equipment for drawing tungsten wire; to lay off workers as necessary, with severance pay allotted; and to resume the making of the earlier nonductile filaments. Fortunately, Mr. Benbow risked making "one more trial" that saved the process. In retrospect, it seems inconceivable that the survival of ductile tungsten could ever have been such a near thing, but the record is clear. Mr. Benbow evidently had not only faith, but courage.

Mr. Benbow in 1914 took a step unusual for a works manager at that time—he sought (and paid for!) help from science. He engaged (the then) Mr. Zay Jeffries, Case School of Applied Science, for consulting and part-time research. Both men profited. Benbow's troubles disappeared, and Jeffries took advantage of a rare scientific opportunity.

Jeffries' major research interest became the effects of changes in grain size and temperature, and of grain deformation, on the properties of metals. His early experience with tungsten sent him in this direction. Two of his important papers[5, 6] extend and explain Dr. Coolidge's tungsten discoveries. In addition, Dr. Jeffries did much to push American metallurgists toward an atomic view of their science, a view attainable by using the more powerful scientific tools as they appeared. The metallurgical microscope was followed in the Cleveland Wire Works by a high-resolution optical spectrograph, and by the first x-ray diffraction equipment used in industry. The Wire Works, now in the Refractory Metals Products Department, soon came to be the Company's center of knowledge about tungsten and molybdenum, calling upon the Research Laboratory only for special help and for transfers of personnel as appropriate.

## DUCTILE TUNGSTEN GOES TO COURT

Ductile tungsten has already given us two first-rate scientific surprises: the Battersea incident and its sequel. A legal surprise at least equally startling is yet in store.

### *General Electric Co. v. Laco-Philips Co.* [7]

Dr. Coolidge's digest of laboratory notebooks shows that he spent 3/29/16 and 3/31/16 at the Mohawk Club, in the historic stockade section of Schenectady, preparing himself to give evidence in two suits in equity being brought by the General Electric Company alleging infringement of Just and Hanaman U.S. Patent 1,081,502 (this suit) and of Coolidge U.S. Patent 1,082,933 (see below). On 5/9/16, Dr. Coolidge records: "Have been testifying for 2 weeks in Laco-Philips suit." This was first tried in the U.S. District Court for the Southern District of New York, Judge Mayer, and resulted in a decree for the complainant. The case then went to the Circuit Court of Appeals, Second Circuit, which affirmed the decree. The General Electric Company had won an important victory in court as the following will show.

Just and Hanaman were among those trying to make a useful tungsten filament in 1904, when they invented the *squirted* filament, the first tungsten filament to be used in a commercial lamp. They squirted powdered tungsten with an organic binder through a diamond die, and got rid of the binder while sintering the tungsten into a fragile filament. The General Electric Company paid them $250,000 for rights to manufacture.[8] The U.S. Pat-

ent Office placed Just and Hanaman in interference with
three other inventors. Just and Hanaman won the interference, and U.S. Patent 1,081,502 was issued to them.
Judge Mayer's decision upholding this patent is an important document in the history of the incandescent lamp
and is of significance also for lamps with filaments of ductile tungsten. The following quotation from the opinion
is taken from pp. 105–106 of Ref. 7:

> I think the invention here in suit is of great merit, en-
> titled to firm support, and second only to Edison's. Edison
> found a dim pathway and transformed it into an illuminated
> road. Just & Hanaman have broadened that road into a
> boulevard, alive with blazing lights. It is not hyperbole to
> say that the tungsten lamp has made millions happier for
> its greater comfort and its better cheer. We accept so read-
> ily as old what was new yesterday that our senses sometimes
> become dull to the real accomplishments of our own time.
> We have almost forgotten the Edison lamp, and as we read
> or ride [sic], watch the play, or sit patiently while men of
> the hour in public places speak in happy or ponderous phrases,
> as we travel by land or water, as we labor on an overcast day
> in the courtroom, the schoolhouse, at the counter, or in
> the heights edging on the modern canyons of a metropolis,
> we find we have a better attitude and a more hopeful tem-
> perament because literally the tungsten lamp has made our
> physical surroundings brighter and more inspiring. To such an
> accomplishment that tribute should be paid which our laws
> contemplate."

The court clearly endorses "Fiat lux" and would no doubt
have rhapsodized about ductile tungsten had the occasion
arisen. The paragraph deserves to be taken seriously, however, because it shows that even in 1916 the brave new
world of industrial research was coming to be taken for
granted.

*General Electric Co. v. Independent Wire Co.*
*Brief for the Plaintiff* [9]

This is the second of the two suits mentioned above, and
the main issue here is the validity of the basic Coolidge
patent on ductile tungsten. The case was tried before
Judge Morris sitting under special assignment in the U.S.
District Court, District of New Jersey. For 5/19/16, Dr.
Coolidge's record reads: "Back from New York, where I've
finished testimony in suit against Independent Lamp and
Wire Company." In this suit, Professor Jeffries, as he was
called in court, was an effective expert witness for the
plaintiff. The brief [9] is of great scientific and historical
value as a record of Dr. Coolidge's work; it supplements and
amplifies the "Finishing Touches" relating to ductile tungsten, some of which will be cited here as FT identified by
date. We are going to begin by using the brief as a scientific monograph, legal questions being deferred. Within
this section, references to the brief will be by page number
in brackets.

As always when complex, imperfectly understood phe-
nomena must be described qualitatively in simple language,
ambiguity and misunderstanding hover in the wings. Perhaps these few warning signposts will help keep them
there. *Working* a metal means changing its shape as by
hammering, swaging, rolling, bending, or drawing through a
die. If no cracking occurs during working, the metal is said
to be *malleable, wrought, pliable, flexible,* or *ductile*; these ad-
jectives are to some extent interchangeable. If cracking
(or fracture) does occur, the metal to that extent is *brittle.* Unfortunately, qualitative descriptions must often be
used when quantitative data are needed. Semiquantita-
tive results can be obtained by making suitable operational

Plate 8.   Dr. E. E. Charlton and Dr. Coolidge shown in the early 1940s with a mobile x-ray machine capable of generating 2-million-volt x-rays, the hardest (most penetrating) then available. Dr. Charlton turned from organic chemistry to electronics, and later to x-rays, after joining the Laboratory in 1920.

tests, as in the case of Table 1 below. One is then likely to find, for example, that all metals are *ductile* to a point depending upon the severity and extent of the test—likewise for *brittle*. And yet we must continue to use these words to tell our story. As a wise and penetrating observer once said, "It is not Nature that is simple; it is we."

To facilitate understanding Dr. Coolidge's work, claims will be selected from the Coolidge patent for brief discussion. From the *coarse powder* claims (9, 11, and 12), we choose Claim 12:

> 12. The process which consists in reducing tungsten oxid to coarse tungsten powder, agglomerating said powder into a coherent porous body, heating said body to drive out impurities and sinter the body, and then subjecting the body to hot mechanical working. [p. 246]

These claims derive from troubles with oxygen carried by tungsten powder in an amount that decreases with the specific surface of the powder; that is, *coarse* powder carries less oxygen than *fine*. Only fine powder could be bought. To make the coarse, Dr. Coolidge had first to make coarse tungstic oxide particles, which he did by a heating (near 1200° C) in fire-clay crucibles; reduction of these particles to coarse tungsten powder followed. See FT 9/11/09, FT 1/5/10, and Comment.

From the *general process claims* (1-8, 10, and 17-23), we choose the following seven [emphasis supplied]:

> 4. The process of producing ductile tungsten wire, which consists in first forming a sintered body of tungsten free from oxygen carbon and other impurities which would render the body unworkable mechanically, then hot swaging such sintered and purified body repeatedly until it becomes

fibrous in structure and then further reducing it by hot drawing. [pp. 228, 229]

6. The process of forming bodies of coherent homogeneous tungsten which consists in agglomerating tungsten powder, sintering the body thus formed, and then subjecting it to hot mechanical working to such an extent as *to improve the internal structure* and to bring the body into a shape adapted for use in the arts. [p. 229]

7. The method which consists in first producing a body of tungsten sintered or compacted throughout and free from such impurities as would render the body unworkable mechanically, and then hot working said body to such an extent as *to deform its crystalline structure* and to bring it into a shape adapted for use in the arts. [p. 229]

8. The method of producing pliable tungsten wire for lamp filaments, which consists in first producing a body of tungsten in a sintered state and free from oxygen carbon and other impurities which would render the body unworkable mechanically and then subjecting this sintered and purified body to mechanical working many times repeated, with gradual reduction in diameter and elongation in length. [p. 229]

17. The method of producing bodies of tungsten pliable and ductile at ordinary or room temperatures from bodies of tungsten which are brittle at ordinary or room temperatures, which consists in heating said brittle bodies to a temperature sufficient to render them susceptible of mechanical working, repeatedly working the bodies while so heated, and during at least the last stages of mechanical working carrying on the working at temperatures such that the mechanical working has the effect of *producing ductility* in the finished product. [p. 230]

19. The method of producing a body of ductile tungsten which consists in repeatedly hot working a tungsten body within a range of temperature in which the body is sufficiently soft to be susceptible of such working, but maintaining the temperatures below those at which the effect of the mechanical working is lost. [p. 230]

23. The method of securing a ductile metal body without previous melting of the metal, which consists in producing from divided metalliferous material a compact body of metal, sintering said body and thereby producing a crystalline body brittle at room temperatures, and then subjecting the body to repeated mechanical working while hot, and continuing the working until the metal remains ductile when cooled. [p. 231]

The claims give little hint of the experimental difficulties that had to be overcome. Swaging near 1500° C had never been done. To do it, Dr. Coolidge—as has been mentioned—was forced to make drastic changes in equipment and technique. The breaking of dies was serious until lubricants for them were found. These matters are mentioned here to make the point, needed later, that the working of tungsten in the making of wire was attended by problems that set it apart from the working of the usual metals, which was based on long experience.

Claim 17 describes the reworking of tungsten that has become spoiled through unplanned re-embrittlement that occurred during the Coolidge process. Claim 23 is of historical value in that it relates to the first large-scale application of what is today called *powder metallurgy*.

A noteworthy feature of the process claims is that they describe pretty well how fine tungsten wire is made today, some sixty years after Dr. Coolidge did his work.

From the *general product claims* (24–29, 32–34), we omit three that are specific to incandescent lamps. The others follow:

24. A wire formed of ductile tungsten. [p. 233]
26. Substantially pure tungsten having ductility and high tensile strength [p. 233]

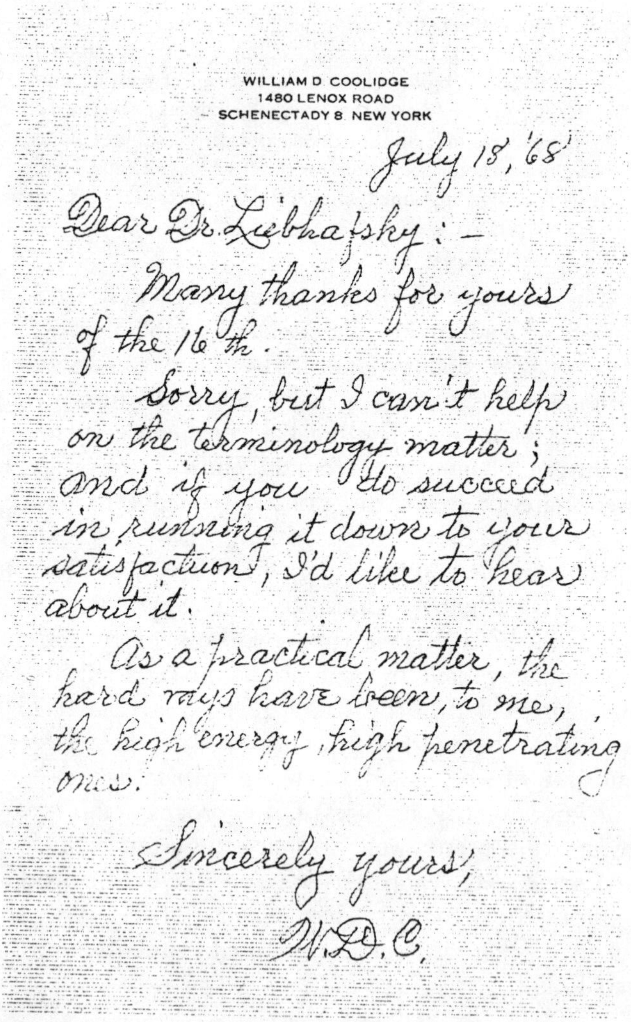

WILLIAM D COOLIDGE
1480 LENOX ROAD
SCHENECTADY 8 NEW YORK

July 18, '68'

Dear Dr. Liebhafsky: —

Many thanks for yours of the 16th.

Sorry, but I can't help on the terminology matter; and if you do succeed in running it down to your satisfaction, I'd like to hear about it.

As a practical matter, the hard rays have been, to me, the high energy, high penetrating ones.

Sincerely yours,

W.D.C.

Plate 9. Holographic reply by Dr. Coolidge to an inquiry about x-rays. With his ninety-fifth birthday approaching, Dr. Coolidge expresses his continuing interest in x-ray terminology.

27. A ductile tungsten wire having a fibrous structure. [p. 233]

28. A form of tungsten metal pliable at room temperature. [p. 233]

33. The material wrought tungsten, having a specific gravity of approximately 19 or greater, and capable of being forged and worked. [p. 233]

34. Wrought tungsten, a solid coherent material characterized by the presence of crystals deformed by mechanical working. [p. 233]

These are sweeping claims. They cover ductile tungsten as a material, no matter how used. Except for Claim 26, which is concerned with "substantially pure" tungsten, they are not composition-of-matter claims: ductility, not composition, is the issue. The keynote of the discussion in the brief is sounded by "We have no fear that any one who has studied the prior art will believe that tungsten is a ductile metal . . ." [p. 109]; here tungsten that has not been put through the Coolidge process is meant. The brief takes the position that Dr. Coolidge *invented* ductile tungsten, a *new* and *useful* material *not obvious* to one skilled in the art [p. 259]; this invention is called *Coolidge metal* [p. 27].

From the *claims relating to offsetting* (13–16, 30 and 31), we select:

13. The process which consists in firing tungsten oxid in a refractory crucible of the Hessian or Battersea type, reducing the oxid to coarse metallic powder, agglomerating the powder into a coherent body, sintering the body, and then subjecting it to hot mechanical working. [p. 236]

16. The process which consists in producing tungsten containing beneficial additions, forming it into a billet, sintering it at high temperature, and mechanically working it a

great number of times at high temperature, and reducing
the temperature during the working. [p. 236]

The first firing of tungstic oxide in a Battersea crucible was done by Mr. Rush, "one of the chemists in the Research Laboratory" [p. 81]. The role played by Hessian crucibles is unexplained in the brief; they do not seem to be
mentioned at all outside these claims, wherein they may
have been included for legal reasons. In any case, the name
"Battersea incident" is historically acceptable. More later.

### Are Brittle and Ductile Tungsten the Same Metal?

This was an important legal question. Let us first try to
find a scientific answer.

In 1910, the predominantly crystalline nature of
metals was not generally accepted. The x-ray diffraction
evidence we have today—thanks largely to Coolidge
tubes—was not yet available. The structure of metals, as
revealed by the metallurgical microscope, was described in
terms of *grains* and *grain boundaries*. The description persists
even though we know today that the grains in a "pure"
metal are almost always single crystals while the grain
boundaries, about which less is known, often contain most
of the impurities.

Professor Jeffries made a valuable forensic contribution
to the General Electric argument by devising an operational test for the ductility of metals and presenting the
results [pp. 144–146]. The test: A lightly clamped wire is
bent forward and backward at room temperature until it
breaks. In one bend, the wire is taken from the vertical

through 90° to the horizontal *and back.* The greater the number of bends, the more ductile the wire. Results for most of the metals tested appear in Table 1.

In an *equiaxed structure,* the grains (crystals) in a metal have axes roughly equal in all three directions.

In *fibrous structures,* the grains (crystals) in a metal have been *elongated,* usually by drawing the metal through a die and usually without grain growth. The long axes of the individual fibers are then roughly parallel to the axis of the wire. To repeat, the fibers are *elongated crystals* produced by deformation. They vary in length. The shorter axes of these elongated crystals are oriented in all possible directions with respect to that of the wire axis. Three kinds of evidence support this view:

Tungsten fibers can be seen under the microscope (Plate 11, p. 66). Tungsten wires severely worked during drawing often split into bundles of fibers "almost like the fibers of a hemp rope" [p. 85]. Finally, the x-ray diffraction patterns clearly show that *preferred orientation* of crystals exists in wires drawn from many metals.[10]

Table 1　*Number of Bends for Wires (7 or 8-mil) of Various Metals*

| Structure[a] | Copper | Iron[b] | Silver | Nickel | Gold | Platinum | Tantalum | Tungsten |
|---|---|---|---|---|---|---|---|---|
| Equiaxed | 14 | 13 | 28 | 13 | 17 | 23 | 16 | 1/9 [c] |
| Fibrous | 6 | 6 | 8 | 5 | 10 | 14 | 8 | 7 |

[a]See text.
[b]Armco iron.
[c]Broke at 20° from vertical before horizontal was reached.

The evidence is clear that all metals can be brought by wire drawing from an *equiaxed* to a *fibrous* structure, the latter being characterized by a preferred orientation of *interlocking elongated crystals.* In the Coolidge process for making tungsten ductile *at room temperature,* all working must be done at a temperature *low* enough to permit lasting deformation (which implies stressing the metal beyond its elastic limit) yet *high* enough so that the ingot, rod, or wire will not break. The proper working temperature fortunately *decreases* with the cross-section of the tungsten. Typically, swaging of the sintered ingot might be begun near 1500° C, and wire drawing concluded near 450° C to give a wire ductile at room temperature. The tensile strength of such wires *increases* as they become thinner, which helps explain why tungsten wire can claim to be the finest that can be drawn, and to be the strongest material known.

The fracture of metals is complex and imperfectly understood, and it is consequently impossible to answer in fundamental terms the question at the head of this section. We can, however, try to see whether the results in Table 1 must be explained by assuming that fibrous (ductile) tungsten—Coolidge metal—is a metal different from equiaxed (brittle) tungsten. We proceed from Professor Jeffries' assumed position that "bendability" is an operational measure of "ductility."

Table 1 shows that three things about tungsten stand out: The ductility of equiaxed tungsten, represented by 1/9, is *phenomenally low.* Tungsten is the only metal for which the table records a lower ductility in the equiaxed state than in the fibrous. The measured ductility of fibrous (ductile) tungsten differs little from that of the other metals.

An interpolation. We could have replaced ductility by brittleness in the previous discussion by taking brittleness to be measured by the reciprocal number of bends, and changing the wording to suit. The point here is that we could use either ductility, or brittleness, or both to describe the behavior of the metals in Table 1.

Without further evidence, it seems reasonable to say that the differences between tungsten and other metals are traceable primarily to the *equiaxed* and *not* to the *fibrous* condition. The changes that occur during wire drawing, which are predominantly *physical,* tend to remove these differences. If anything, equiaxed (brittle) and not fibrous (ductile) tungsten is anomalous. Scientifically, there is no compulsion to regard ductile tungsten— remarkable though it is—as a new metal: we may say that Nature flawed tungsten by making it excessively brittle and that Dr. Coolidge repaired the flaw.

Molybdenum resembles tungsten more closely than does any other metal in Table 1. Molybdenum might well have been given the Jeffries test [see p. 285].

### Notes about People

The following quotations from the brief add to the story of ductile tungsten.

Appraisal by counsel of Professor Jeffries and Dr. Coolidge:

> His [Professor Jeffries'] testimony shows a knowledge of the subject so extended and so profound that we have no hesitation in presenting him to the court as the most skilled

Plate 10.   The directorial staff of the Research Laboratory gathers in 1952 for a postretirement photograph in the Laboratory's new home at the Knolls outside Schenectady. From the left: Willis R. Whitney, Irving Langmuir, William D. Coolidge, Saul Dushman, and Albert W. Hull. In 1932 Dr. Coolidge succeeded Dr. Whitney as Director; both men were Vice Presidents of the General Electric Company before they retired. The others had subordinate titles. The chance seems negligible that a fledgling industrial laboratory could have attracted five such outstanding men on the basis of statistics alone.

expert in the metallurgy of tungsten, not even excepting Dr. Coolidge, who made the art. [p. 26]

## Dr. Coolidge as a metallurgist:

The metallurgists were so steeped in the theories of their art that they were constitutionally incapable of even attacking the problem along Coolidge lines; if Coolidge had been as expert a metallurgist as he was a chemist, he too, it is probable, would have been misled as the others were misled and would have failed as they failed, not only to accomplish anything worth while but even to see clearly what it was that was to be accomplished. It was the happy accident that a man of Dr. Coolidge's deep physical and chemical training and of his inventive genius, not too thoroughly saturated with the limitations and theories of the metallurgists, was attracted to this metallurgical problem, that made its solution even thinkable. [p. 274]

## Dr. Coolidge says of his interview with Dr. Blau (FT 11/8/09):

I remember this circumstance very well because of the excitement and surprise and incredulity which he manifested at the time. He asked me over and over again what it was. I told him that it was pure tungsten wire, only to have his question repeated again and again. [p. 55]

## Dr. Blau speaks of the Coolidge process for making ductile tungsten:

Now, even after I know the procedure quite exactly, I can say that I have never met with any other substance whose physical properties are so gradually, fundamentally changed by a process of mechanical working as is the case with tungsten. The first steps of the mechanical working cause such an inappreciable change in the brittleness of the tungsten, that

> I found it almost incomprehensible how anyone could place
> such faith in a hoped-for result, and could show such pa-
> tience in the carrying through of an apparently hopeless ex-
> periment as has been shown in the case in question by the
> engineers and chemists of the General Electric Company, for
> after the tungsten has been entirely changed in form by
> mechanical working at a red heat, and has been reduced even
> to a third of its original cross-section, it still shows no
> appreciable decrease in its brittleness. [pp. 54, 55]

By this time, Dr. Blau's employer was making lamps under
license from the General Electric Company.

### More about Battersea

The "something" transferred to tungstic oxide during
firing in a crucible (Battersea or Hessian) in all likelihood
was never completely identified. The patent has this to
say:

> Starting with tungstic oxid ($WO_3$), as pure as commercial-
> ly obtainable, I dissolve it in ammonia water and thus form
> ammonium tungstate, which is purified by crystallization
> over a steam bath and is then, after washing by distilled wa-
> ter and drying, decomposed by the application of moderate
> heat, say 400 to 500 degrees C. which drives off the ammon-
> ia and leaves purified tungstic oxid. The tungstic oxid thus
> purified is heated in a gas furnace for about five hours in a
> covered Hessian or Battersea crucible of the ordinary and
> well known type, after which step it is found to be coarser
> and of higher density and to contain matter, probably vola-
> tilized from the crucible, the effect of which, as will be here-
> inafter explained, is beneficial. The substances thus added
> to the oxid of tungsten, which for convenience I may desig-

nate as additions, consist largely of alumina, and silica. I have obtained best results when these additions occurred in the proportion of from 0.8 to 1.5 of one per cent.

If on analysis it is found that less than say 0.8 of one per cent of additions is present, the operation of firing in the Hessian or Battersea crucible is repeated as many times as may be necessary, a new crucible being used each time. The material, which has somewhat melted and agglomerated into a mass, is ground after each firing. [Coolidge, U.S. Patent 1,082,933, p. 2].

This method of doping testifies to the urgency and the baffling nature of the offsetting problem. Battersea crucibles were expendable! No matter how many it took, adequate "beneficial additions" had to be given a sound experimental foundation. That a better way was then found is shown by the following quotation, which explains Claim 16 (see FT 6/21/10):

I have succeeded in preventing offsetting in drawn tungsten filaments without using the Battersea crucible process by mixing with the tungsten powder certain refractory materials, such as the oxids of thorium, zirconium, yttrium, erbium, didymium, or ytterbium. I have used with especial success thorium nitrate, which gives thoria when decomposed. [p. 91]

Offsetting is easily seen on an enlarged photograph, and the beneficial effect of thoria can be shown in the same way; see Plate 11 (p. 66).

Offsetting results from objectionable grain growth in straight-line filaments during operation with alternating current at the high temperatures needed for high efficiency. This grain growth produces grain boundaries that extend across and through the filament. Lateral movement

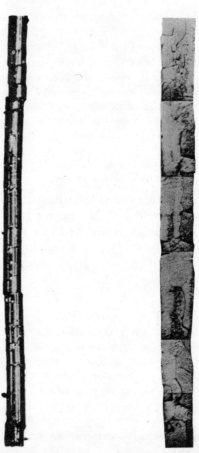

Plate 11. Offsetting and offsetting prevented: photomicrographs sub-
mitted to Judge Morris and taken from Ref. 9, Chapter 2. Catastro-
phic offsetting, produced by operation on alternating current, is evi-
dent along the vertical boundaries of the straight-line filament on the
left. The filament on the right has been "doped" with thorium oxide
and comparably operated. Here the vertical boundaries show no offsetting, although long tungsten grains (crystals), some of which extend
across the horizontal lines, can be seen. The lines are artifacts, being
sectional boundaries in this composite photograph.

of successive lengths of filament results in offsetting and early failure. In a drawn wire, added thoria appears in long threads parallel to the axis of the filament [p. 241], and these threads are thought to obstruct grain growth in the filament, preferentially in radial directions. The objectionable grain boundaries along which offsetting occurs are consequently thought not to form.

The prevention of offsetting led to a rapid transition of ductile tungsten from the Research Laboratory:

> We have seen also that at a later date Dr. Coolidge made some wire from ingots formed of tungsten which had been baked in a Battersea crucible; filaments made from this wire did not "offset"; thereupon the Harrison factory changed over as fast as it could from the *squirting* process to the *drawn wire* process at an expense of approximately a million dollars, and the whole world, including the Defendant, has changed over in the same way, just as fast as it could get the information and the apparatus. The drawn wire was a great invention, but lamps which could not have been burned successfully on alternating current would have been useless on many commercial circuits, so in the improvement which adapted the Coolidge lamps to alternating current work we again have a new process and a new result of great value. [p. 238]

The transition was costly. Apparatus to the estimated value of $500,000 for making squirted tungsten filaments had to be scrapped. An inventory loss of $400,000 on lamps with these obsolete filaments was incurred [p. 29]. The cost of Dr. Coolidge's research ($116,000) was small by comparison.

Langmuir's work soon made straight-line filaments obsolete. His gas-filled lamps contained coiled tungsten filaments; to make these, ductility in wire was all-important. In them, objectionable grain growth led to sagging, which (as Dr. Alader Pacz in Cleveland discovered) could be eliminated by doping with alkali silicates to produce a microstructure of elongated interlocking tungsten crystals; see Chapter 1, Ref. 5b, p. 226. The "something" transferred in the Battersea incident probably included a little alkali metal along with alumina and silica. If so, Dr. Pacz's important discovery becomes a logical corollary of the Battersea incident and makes a closed loop of the improbable chain of events listed in Chapter 1.

Much more research could be done on ductile tungsten. Two recent papers[11, 12] show that annealing of doped lamp-filament tungsten causes rows of voids to form along the direction of the filament axis. These voids are attributed to the vaporization of the doping agents or of their reaction products, reminiscent of the threads of thoria mentioned above. The wonder grows that the Battersea incident and its sequel could occur.

For completeness, we record without discussion that the brief cites and discusses prior art as disclosing the addition of thoria to tungsten, but concludes it to be inapplicable.

## General Electric Co. v. Independent Lamp and Wire Co. The Decision[13]

Judge Morris, sitting under special assignment as was mentioned above, held valid *all 34 claims* of the Coolidge patent on June 29, 1920, over two years after the date of plaintiff's brief on final hearing. His opinion sanctions the use of "Coolidge metal" to describe wrought tungsten. He considered wrought (ductile) tungsten to be a patentable invention, which means that he regarded brittle and ductile tungsten as *different metals.*

The General Electric Company had thus won a sweeping legal victory even more impressive than that in the Laco-Philips case. Its U.S. patent position with respect to tungsten-filament lamps and to ductile tungsten seemed impregnable.

## *General Electric Co. v. De Forest Radio Co. et al.* [14]
### *The Legal Surprise*

The Coolidge patent had not been tested with respect to ductile-tungsten electron-emitters used, for example, in radio tubes. To make this test, the General Electric Company filed suit charging infringement resulting from "the manufacture and sale by the De Forest Company of radio tubes having ductile tungsten filaments made to its order by Mallory Company, an alleged infringing manufacturer of drawn tungsten wire, and the resale of those tubes by Robelen Piano Company, a De Forest distributor" [Ref. 14, p. 91]. The suit was brought in the U.S. District Court, Delaware District, before Judge Morris, whom we met in New Jersey.

The following eight claims were excluded as inapplicable to radio tubes: Claims 8, 30–32 (lamp filaments); Claims 13 and 15 (Battersea firing); Claims 25 and 29 (incandescent lamps). Many of the remaining twenty-six claims are vital for *all applications of ductile tungsten.* On January 15, 1927, Judge Morris held all twenty-six claims *invalid* and dismissed the suit. The heart was thus cut out of the Coolidge patent. A legal surprise indeed!

Judge Morris reversed himself in a fifteen-page opinion because of new specific defenses, because of new evidence, and because of a decision of the House of Lords upon the corresponding British Coolidge patent, this decision having been made since the Coolidge patent was held valid in New Jersey. The Delaware opinion reflects much soul-searching and could not have been easy to write. It cannot be discussed adequately here, but is recommended as absorbing reading for anyone interested in relationships between research and the law. The main points follow. Bracketed page references are to the opinion.

Tungsten is *inherently ductile even at room temperature* and hence unpatentable as a product because "a discovery of properties is not a manufacture" [p. 96]. One supporting argument is that the "workableness" (ductility) of tungsten hot must be inherent, or the metal could not be worked at all; and that there is no stage in the Coolidge process at which this inherent property is replaced by an invented one, the result being that *inherent ductility* persists in tungsten even to room temperature. The argument is supported by the established ductility at room temperature of single-crystal tungsten that has not been put through the Coolidge process. Dr. Coolidge had demonstrated this ductility, but had taken the position that "single grains are laboratory and museum curiosities." The defendant's expert witness, Dr. Wadsworth, was able to draw a single crystal in the form of a filament through five dies, increasing its length from 7-7/8 to 12 inches, and reducing its diameter from 4.5 to 3.4 mils. That he could not repeat the experiment with a like crystal did not matter: one single crystal had been found ductile; therefore, according to the court, tungsten was inherently ductile. The legal position is that ductility is *inherent* in *both* tungsten samples in Table 1, as it is in all other metals of the table. "Coolidge metal" is *not* a new metal: it is a discovery but not a patentable invention. Ductility in tungsten cannot be *produced* by any means, no matter how difficult or ingenious: ductility is there all the time. Cf. italicized portion of Claim 17 given above.

As regards conception, which is necessary to invention, Judge Morris says:

> Drawn tungsten was conceived of and recognized by others prior to the earliest possible effective date of the patent in suit as a means for better satisfying the want that existed in the incandescent lamp industry. [pp. 101, 102]

On the process claims, Judge Morris takes this position:

> . . . a properly prepared body of tungsten has many properties in common with other metals. Other metals may be worked; so may tungsten. Each has its own equiaxing temperature, sometimes called its annealing temperature, at and above which its grains, when deformed by mechanical forces, promptly return, upon the removal of the force, to their equiaxial state, and below which grains, deformed by the application of mechanical force, remain permanently in their deformed condition. . . . Laying aside its equiaxing temperature—a matter of discovery—I find nothing in the method of working tungsten that differs in principle from the working of other metals save, perhaps, that the force of adhesion between its grains is not great, and that consequently it must be worked more gently and by easier stages. But this, too, is an inherent, natural property of the metal. [p. 103]

Judge Morris goes on to say that the failure to draw tungsten before this century is attributable primarily to the *lack of need* for incandescent-lamp filaments, and that this

Plate 12. The line of succession completed. Dr. Coolidge, behind the cake commemorating his one hundredth birthday, is flanked by Dr. C. Guy Suits, his successor (right), and by Dr. Arthur M. Bueche, who succeeded Dr. Suits in 1965. Dr. Bueche's title is Vice President, Corporate Research and Development, and the Research Laboratory is now part of the Corporate Research and Development Center. The cake is illuminated by one hundred barely visible miniature incandescent lamps with tungsten filaments. The center column, of tungsten carbide, supports a one-third carat gem diamond, man-made at the Research Laboratory.

failure is consequently of little *legal* significance. [pp. 103, 104]

Judge Morris appreciated that there need be no correlation between the patentability of a discovery, its scientific merit, and its importance:

> But, by saying that I see no merit in this and other claims, I intend to convey only the thought that they are not within the scope of the patent laws. I do not mean to be understood as indicating that the discoveries involved in making drawn tungsten ductile cold were not of a character entitling the discoverer to the highest credit. That they have greatly served mankind is established by the efficiency of drawn tungsten and the universality of its use. But this fact is sufficient to turn the scale in favor of invention only when the other facts in the case leave the question in doubt. [p. 104]

The defendants presented as one defense that Dr. Colin G. Fink, and not Dr. Coolidge, was the inventor, "if invention there was," of both process and product. The argument was that Dr. Fink learned how to draw molybdenum wire while Dr. Coolidge was in Europe in 1908; that Dr. Coolidge had the benefit of Fink's interim experience upon his return in September; that molybdenum resembles tungsten as regards wire drawing; and that tungsten was drawn successfully on October 21, 1908, by methods that worked with molybdenum. Judge Morris found it unnecessary to consider this defense. Note that Dr. Coolidge mentions Dr. Fink's activity in FT 10/21/08, that he acknowledges the help of his colleagues in his paper on ductile tungsten[15] as quoted under FT, and that Dr. Coolidge is mentioned by name in a contemporaneous acknowledgment[16] by Dr. Fink.

The decision also cites prior art to invalidate Coolidge claims of adding thoria to prevent offsetting. As similar prior art is discussed in the International Lamp Co. brief (see above), and as thoria is no longer added for this purpose, we shall omit discussion here.

The matter of credit for Dr. Fink would not have been mentioned after all these years were it not an example of a difficulty common to much successful research—particularly to successful industrial research. Failures give no such trouble. The difficulty can be serious; it involves people; it often defies rational analysis. There is no doubt in my mind that Dr. Coolidge deserves all the recognition he has received for making tungsten ductile.

Judge Morris' reversal of himself led Dr. Coolidge to decline the Edison Medal (see FT 12/15/26 and 12/15/27). He wrote to the AIEE in part:

> Judge Morris has just handed down a decision to the effect that my ductile tungsten patent is invalid. This decision, coming from a man of Judge Morris' standing, proves to me that the best of men could question my right to the Edison Medal. My appreciation of that great pioneer, Mr. Edison, in whose honor the medal was established, and my admiration for its former recipients are such that I would not for the world do anything that could in any way detract from the luster of that medal, which should stand for generations to come as one of the most coveted prizes for meritorious work in the electrical field.[17]

Coolidge Patent 1,082,933 dealt with what is in my opinion the most important and successful project carried through in the Research Laboratory. It was struck down fourteen years after its issue. One is haunted by the feeling that ductile tungsten is nevertheless an excellent example of what the founding fathers must have had in mind when they called for a patent system "to promote the Progress of Science and useful Arts" in Article 1, Section 8 of the Constitution, and that most of the Congressmen who approved the many acts relating to patents would have been surprised to see the Coolidge patent struck down. Yet, Judge Morris' reversal is easy to understand. This case warrants expert study in connection with any examination of the future usefulness of the patent system in these days of growing complexity. What proportion of patents could survive appeal to the circuit courts? Where can we find enough judges capable of evaluating opposing arguments in complex cases? How many years must one wait until the law has had its say? How predictable is the result? What will it all cost?

## CONCLUSION

Here ends this incomplete account of what happened to ductile tungsten after it left the Research Laboratory. Present incandescent lamps are much improved over those of 1912, but their mainstay is still a filament of ductile tungsten made by essentially the Coolidge process. No other product in the history of organized industrial research has a better record.

That record includes not only Coolidge tubes and other vacuum tubes already mentioned but also such applications as springs, furnace windings, tungsten-glass seals, and electrodes for atomic-hydrogen welding and for other purposes.

The history of ductile tungsten in the courts shows

once again that research men need the help of lawyers who understand science. It reveals that legal briefs are valuable as historical sources provided one remembers that a brief must necessarily make the best possible case for one party because the adversary method operates in court.

# EPILOGUE

## Love's Labor Lost?

By no means! The chance to honor Dr. Coolidge at this time is reward enought for writing this book. After all, what other centenarian has directed a large research laboratory and made two great contributions to modern civilization?

The author hopes for more. Perhaps the book will make tungsten better known, as it deserves to be, and not for scientific reasons alone. Of course, the preface has shown that tungsten can match superlatives with any other metal—even with platinum, which also fills vital and unusual needs in our civilization though it is much less abundant than tungsten, of which no scarcity is foreseeable. China alone has high-grade reserves that can yield some 10 billion pounds of tungstic oxide! The discovery of these reserves makes a tale as dramatic as does the conquering of tungsten's brittleness, or the Battersea incident, or the fate of the Coolidge patent in the courts. For once, it is a tungsten tale in which General Electric has no part. The Chinese tungsten reserves were discovered in 1911 by Mr. K. C. Li,[1] who was then prospecting for tin (not tea!) in China. To expedite an evening meal in the month of May, he collaborated with an innkeeper's teenage daughter to keep a fire going in the stove that her father had built some twenty-five years earlier from a mineral that threatened to demolish his garden when chunks of it came rolling unpredictably down a neighboring hill. Mr. Li found that the mineral was wolframite, the richest of all tungsten ores. Alas, Mars prevailed over Cupid. World War I made China an important exporter of the mineral, and in 1936 Germany quietly began stockpiling it for World War II: tungsten diplomacy antedates that now being practiced by the Arabs with their oil.

Why not a television series to tell the complete tungsten story, which now includes even "the biggest sweet
spot in golf"?[2]

But, to return to this book. It will earn its most
important bonus if it leads to a better understanding of
research—especially of organized industrial applied research,
whose first major success it describes. Here an investment
of $116,000 brought a spectacular return. Is this as likely
today? How did the research by Dr. Coolidge differ from
that which led to the double helix?[3] Questions such as
these have come to concern us all. They do not often have
simple answers. And, when answers cannot be agreed upon,
a man must make his own.

In retrospect, the ductile tungsten project was a research director's dream. Why? It was undertaken to fill a
great need; namely, more light at tolerable cost. It
showed early promise of ultimate realizability as it was
known that tungsten ductile at room temperature would
surpass all other filament materials, and as tungsten was
known to be ductile at high temperature. It presented a
problem difficult enough to test to the limit the best
of research laboratories, the problem being to extend the
high-temperature ductility to low temperature; difficult,
yes, but still a problem clearly defined. The promise, enig-
matic like that of Mona Lisa's smile, became alluring in
1908 because the General Electric Company had laid out
$760,000 to buy[4a] American rights to four European pa-
tents for filaments known to fall short of being satisfactory. This was over six times as much[4b] as the Research
Laboratory spent on tungsten-filament research in the
period 1906–1911. Need, promise, problem—all three criteria
seemed favorable for ductile tungsten. The project was of

a kind completely in tune with the old General Electric slogan, "More goods for more people at less cost," which is never without appeal.

Much continues to be written about research and what it means to our civilization, and much of this writing appears to be done without Montaigne's hard question "Que sais-je?" firmly in mind. The story of ductile tungsten, even told incompletely as it is here, can help keep such writing down to earth. The siren tones of Don Quixote are not enough: Sancho Panza also must be given his say. He would agree that the "need-promise-problem" analysis is useful even if it does no more than sort out projects with which the analysis cannot cope owing to their complexity, our ignorance, or their seeming hopelessness.

In this century, the electrical industry has benefited enormously from research on a small number of chemical elements. In the ductile tungsten story, tungsten has the center of the stage with molybdenum in a supporting role. Carbon is turned into man-made diamonds. Nuclear energy relies on uranium. Silicon for semi-conductors is supported by germanium (and by certain compounds that need not be considered here). To each of these cases, the "need-promise-problem" analysis and the old General Electric slogan both seem applicable; the reader may wish to see for himself. One naturally wonders whether the periodic table has further projects of this kind to offer. Is the horn of plenty running low?

In all these cases, research successfully played the role of Lady Bountiful in accord with the old General Electric slogan given above. The role becomes more difficult to play as discoveries multiply, and it has come to be denigrated by many who think they are "telling it like it is" the while they "do their own thing." They preach antimaterialism that is usually blind because it does not strike a balance between good and harm. These attacks, though they seem to be diminishing, are serious to the extent that they involve *la trahison des clercs,* scientists included. The attackers ought to consider the net good done by ductile tungsten as one reason for not throwing out the baby with the bath.

But research faces troubles more serious than disenchantment with Lady Bountiful. Our needs today—such as cures for cancer, acceptable sources of abundant energy, abatement of pollution—are not needs that Lady Bountiful can meet. Research is increasingly being called upon to rescue civilization—not to embellish it. A comparison of Lady Bountiful's slogan with *"Enough* energy *always* for *all* people at *minimum cost"* drives home the difference: for example, in spite of all that research can do, the minimum cost may be more than all people are willing or able to pay. Many Americans, not all of them Archie Bunkers, have come to expect that research should fulfill the World War II slogan: "The difficult we do today; the impossible takes a little longer." Again, a "need-promise-problem" analysis is in order. No question about the need, but can one see reasonable promise, and what are the problems first to be solved? We may be confronted here by massive misconceptions and unrealistic expectations held by many, among them Congressmen who vote millions or billions for research. An acquaintance with Dr. Coolidge and his work might for some be the beginning of wisdom, especially if the acquaintance is tempered by the realization that success such as his is the exception—not the rule. No one contends that such knowledge will lift our troubles, but no more now than in Lincoln's time can we escape history—let us learn more about the successes and the failures of research!

# Notes and References

## Chapter 1

1. The history of the General Electric Research Laboratory is well documented. Most of the works listed in chronological order below contain information about Dr. Coolidge. Throughout this book, reference to these works is by the bracketed acronyms shown.

a. J. T. Broderick, *Willis Rodney Whitney*, Fort Orange Press, Albany, 1945. [WRW]

b. L. A. Hawkins, *Adventures Into the Unknown*, William Morrow and Company, 1950. [AIU]

c. K. A. Birr, *Pioneering in Industrial Research: The Story of the General Electric Research Laboratory*, Public Affairs Press, Washington, 1957. [GERL]

d. *The Collected Works of Irving Langmuir*, Guy Suits, Gen. Ed., Pergamon Press, New York, 1961-1962. Twelve volumes. [IL]

e. J. A. Miller, *Yankee Scientist, William David Coolidge*, Mohawk Development Service, Schenectady, 1963. [YS]

YS is the best single source of information about Dr. Coolidge; but note also the admirable sketch in AIU, Chapter 5; and WRW, pp. 86 and 87.

It is fortunate that this documentation exists. Dr. Coolidge is not listed in the index of W. C. Dampier, *A History of Science and Its Relations with Philosophy and Religion*, Cambridge University Press, London, 4th ed., 1966; nor in J. D. Bernal, *Science in History*, Hawthorn Books, New York, 3rd ed., 1965. On p. 518 of the second book, we find: "It was in 1909 that the General Electric Company chose the already distinguished physicist Irving Langmuir to direct their new research laboratory." When Dr. Langmuir joined the Research Laboratory as a young man in 1909, it was already a going concern, which he was never to direct.

2.   Of defining "research" there is no end. Here, research is the quest for new (not merely more precise) knowledge. If the quest has a formulatable practical objective, the research is applied. Otherwise, it is pure. "Industrial research" tells only where the research is done. Thus it can be of either kind, but applied research is usually dominant, as it was with Dr. Coolidge.

3.   O. W. Holmes, Sr., *Elsie Venner*, Houghton Mifflin and Co., New York, 1892, Chapter 1.

4.   H. A. Liebhafsky, H. G. Pfeiffer, E. H. Winslow, and P. D. Zemany, *X-ray Absorption and Emission in Analytical Chemistry*, John Wiley and Sons, New York, 1960. The dedicatory sentence of the book is quoted.

5.   (a) C. J. Smithells, *Tungsten*, Chemical Publishing Co., New York, 1953. (b) K. C. Li and C. Y. Wang, *Tungsten*, Reinhold, New York, 1955, contains an excellent description of the manufacture of ductile tungsten on pp. 215–228.

6.   YS, p. 67; Ref. 5b, p. 224.

7.   YS, p. 70.

8.   AIU contains admirable sketches also of Whitney and Langmuir.

9.   O. W. Holmes, Jr., *Collected Legal Papers*, Harcourt, Brace, Jovanovich, New York, 1920, p. 266.

10.   IL, v. 12, p. 69.

11.   YS, p. 75.

12.   IL, v. 3, p. 234. Important papers on related work, with K. H. Kingdon as coauthor, will be found near the paper cited.

13.   GERL, pp. 120 and 121. The effect on the Research Laboratory of the 1937 depression was similar, but quicker and shorter. HAL, personal knowledge.

14.  HAL, personal knowledge.

15.  WDC to ALM, letter dated June 9, 1932.

16.  K. H. Kingdon and H. C. Pollock, *Research Laboratory Bulletin*, General Electric Co., Summer 1965, p. 9. The last two quotations are from Dr. Kingdon's notebook.

17.  K. H. Kingdon, H. C. Pollock, E. T. Booth and J. R. Dunning, *Phys. Rev.*, **57,** 749 (1940).

18.  AIU, p. 104; GERL, p. 139; YS, p. 158.

19.  YS, p. 174, *et seq.*

20.  AIU, p. 5.

21.  Heraclitus, with the sensitivity for change shown by this quotation, had at least one characteristic of a good research director.

22.  W. D. Coolidge, *Armour Research Engineer and Alumnus*, **3,** no. 2, 9–11, 38 (December 1937).

23.  W. D. Coolidge, *Trans. AIEE*, **29,** Part 2, 961 (1910).

24.  W. D. Coolidge, *Trans. AIEE*, **31,** Part 1, 1219 (1912).

25.  AIU, p. 29.

26.  YS, pp. 26, 27, 37, 39.

27.  IL, v. 3, p. 3.

28.  W. D. Coolidge, *Phys. Rev.*, **2,** 409 (1913).

29.  AIU, pp. 65–69.

30.  H. A. Liebhafsky, H. G. Pfeiffer, E. H. Winslow, and P. D. Zemany, *X-rays, Electrons and Analytical Chemistry*, John Wiley and Sons, New York, 1972.

31.  YS, p. 173.

32.  YS, p. 65.

33.  AIU, p. 29.

34.  GERL, p. 37.

35.  YS, pp. 144, 145.

36.  Dr. Coolidge, as quoted in YS, pp. 69, 70.

37.  AIU, pp. 31, 32.

38.  AIU, p. 24.

39.  YS, pp. 117, 119.

40.  E. G. Rochow and W. F. Gilliam, *J. Amer. Chem. Soc.*, **61,** 3591 (1939).

41.   AIU, p. 33.

42.   YS, p. 134.

43.   YS, p. 80.

44.   The Brahmins of Ref. 3. The young Coolidge by contrast was an example of the "Nature's republicanism" invoked by the elder Holmes, but an example who was definitely not "a large uncombed youth." See Ref. 3, p. 5.

45.   O. W. Holmes, Jr., *The Common Law*, Belknap Press of Harvard University Press, Cambridge, Mass., 1963, p. 5.

## *Chapter 2*

1.   See "Life and Light Facts" on the wrappers of General Electric Soft-white 100-watt Bulbs, 70c for a package of two. Note that filament temperature increases with "Actual Socket Voltage," and that lamp life concomitantly decreases.

2.   Chapter 1, Ref. 5b, p. 392.

3.   W. D. Mogerman, *Zay Jeffries*, American Society for Metals, Metals Park, Ohio, 1973.

4.   GERL, p. 39..

5.   Z. Jeffries, "The Metallography of Tungsten," *Trans. AIME*, **60,** 588 (1919).

6.   Z. Jeffries and P. Tarasov, "Tungsten and Thoria," *Proc. Inst. Met. Div., Trans. AIME*, **77a,** 395 (1927).

7.   233 Fed. 96 (Circuit Court of Appeals, Second Circuit, 1916).

8.   GERL, p. 38.

9.   *Brief for Plaintiff on Final Hearing:* Suit in Equity No. 648 on Coolidge Patent No. 1,082,933, December 30, 1913; Howson and Howson, Solicitors for Plaintiff; March 1918.

10.   H. P. Klug and L. E. Alexander, *X-ray Diffraction Procedures*, John Wiley and Sons, New York, 1954, p. 555 *et seq.* See also Chapter 1, Ref. 5a, p. 156 and Figs. 156–158.

11.   R. C. Koo, *Trans. Met. Soc. AIME*, **239,** 1996 (1967).

12. G. Das and V. Radcliffe, *ibid.*, **242**, 2191 (1968).

13. 267 Fed. 824 (D.N.J. 1920).

14. 17 Fed. 2d (D. Del. 1927).

15. Chapter 1, Ref. 23.

16. C. G. Fink, *Trans. Am. Electrochem. Soc.*, **17**, 229 (1910).

17. YS, p. 117.

## *Epilogue*

1. Chapter 1, Ref. 5b, Foreword. Required reading for anyone interested in tungsten.

2. *Golf Digest*, Golf Digest, Inc., Norwalk, Conn., January 1974, Acushnet advertisement on inside cover.

3. J. D. Watson, *The Double Helix*, Atheneum Press, New York, 1968. Pure research, vividly described.

4. GERL: (a) p. 38; (b) p. 39.

# APPENDIX

## William David Coolidge

*Medals and Awards*

1914　*Rumford Medal* of the American Academy of Arts and Science for his invention of ductile tungsten.

1926　*Howard N. Potts Medal* of the Franklin Institute for the originality and ingenuity shown in the development of a vacuum tube that has simplified and revolutionized the production of x-rays.

1926　*Louis Edward Levy Gold Medal* of the Franklin Institute for his paper on "The Production of High Voltage Cathode Rays Outside the Generating Tube."

1927　*Gold Medal* of the American College of Radiology for his contribution to radiology and the science of medicine.

1927　*Hughes Medal* of the Royal Society, London for his work on the x-rays and the development of highly efficient apparatus for their production.

1927  *Edison Medal* of the American Institute of Electrical Engineers for his contributions to the incandescent electric lighting and the x-ray arts.

1932  *Washington Award* of the Western Society of Engineers for devoted, unselfish, and preeminent service in advancing human progress.

1937  *John Scott Award* granted by the City Trusts of the City of Philadelphia for applying a new principle in x-ray tubes.

1939  *Faraday Medal* of the Institution of Electrical Engineers of England for notable scientific or industrial achievement in Electrical Engineering.

1940  *Modern Pioneer Award* of the National Manufacturer's Association.

1942  *Duddell Medal* of the Physical Society of England for his invention of the Coolidge x-ray tube.

1942  *Orden al Merito* of the Chilean Government for his many services to civilization.

1944  *Franklin Medal* of the Franklin Institute for his contributions to the welfare of humanity, especially in the field of the manufacture of ductile tungsten and in the field of improved apparatus for the production and control of x-rays.

1951  (The first) *K. C. Li Gold Medal, for the Advancement of the Science of Tungsten* by Columbia University.

1963  *Roentgen Medal* for contributions to the science and technology of x-rays.

1972  *Power-Life Award* by Power Engineering Society of the IEEE for his contributions to the science of x-rays, the medical profession, and the welfare of humanity.

1973  *Climax Molybdenum Wedgwood Medallion* for pioneering work leading to the invention of ductile tungsten and molybenum.

1973  *William D. Coolidge Award* by the American Association of Physicists in Medicine.

Dr. Coolidge has also received numerous honorary degrees and is an honorary member of many organizations. His *United States* patents number 83, the first being No. 935,463, *Dies and Die Supports*, 1909; and his last, No. 2,181,724, *Electrostatic Machines*, 1935. All are assigned to the General Electric Company.

# Index